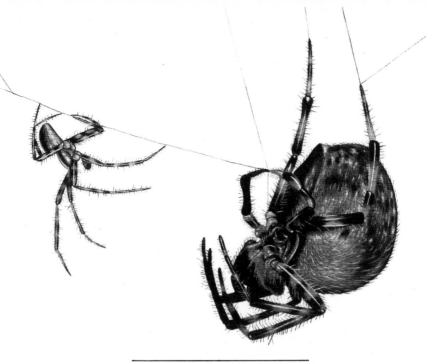

LIFE STORY

SPIDER

MICHAEL CHINERY

Photographs by
Barrie Watts

Illustrated by
Alan Male

SCHOLASTIC INC.

New York   Toronto   London   Auckland   Sydney
Mexico City   New Delhi   Hong Kong   Buenos Aires

ISBN 0-439-70207-0

12 11 10 9 8 7 6 5 4 3 2 1   4 5 6 7 8 9/0

Printed in the U.S.A.   23

First Scholastic printing, October 2004

Designed by James Marks

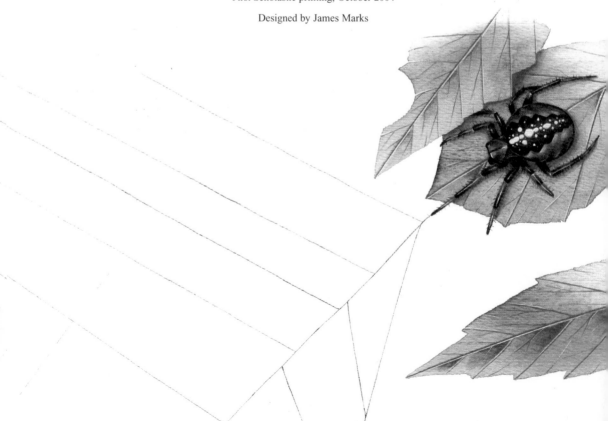

INTRODUCTION

Spiders are some of the very few animals that make traps to catch their food. This book will show you how the garden spider makes its silken web, and you can discover why the spider doesn't get caught in its own web. You can also find out what happens to the insects that are trapped in the web. You will learn that the female spider wraps hundreds of eggs in silk to keep them safe through the winter, and find out how the baby spiders make silken "parachutes" which carry them to new homes.

Not many people like spiders, but they are really very interesting animals. They are useful, too, because they catch flies and other annoying insects. Some spiders chase their prey, but many others make webs to trap the insects.

The garden spider pictured here is one of the most common web-spinners. You can see its wheel-shaped webs on bushes and fences almost everywhere.

4

Look at its eyes. The garden spider actually has eight eyes, but it still can't see very well. It relies on what it feels to tell it what is happening around it. When an insect flies into the web, the spider feels the web vibrate. That's how the spider knows that food is near.

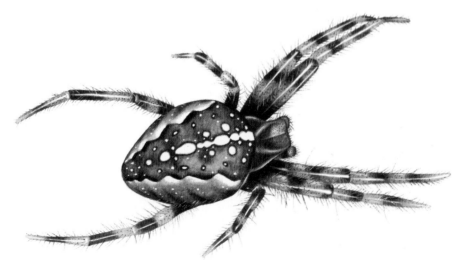

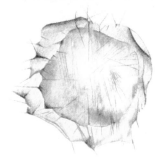

The female garden spider lays her eggs in the autumn and wraps them in a bundle of silk threads. The silk comes from glands at the rear of her body. Silk is very important to spiders and they make several kinds of silk to use for different purposes.

The bundle of silk containing the eggs is often fixed in a gap in the bark of a tree or under a windowsill. It stays there all through the winter.

The eggs are yellow at first, but they gradually turn gray as the baby spiders grow inside them. The eggs in the picture are almost ready to hatch.

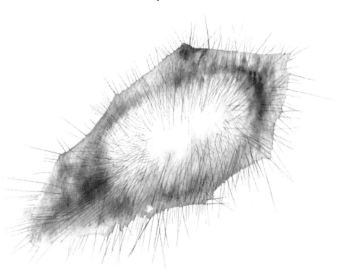

The garden spider's eggs hatch in the spring. The brightly colored baby spiders, called spiderlings, struggle out of their silken ball and trail silk wherever they go. You can see the silken strands very clearly in the photograph.

The spiderlings cluster together for a day or two. There may be several hundred of them, but not many of them will survive. They soon get hungry and start to eat each other.

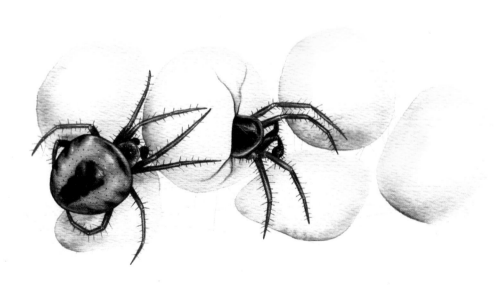

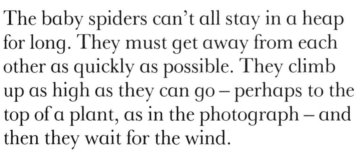

The baby spiders can't all stay in a heap
for long. They must get away from each
other as quickly as possible. They climb
up as high as they can go – perhaps to the
top of a plant, as in the photograph – and
then they wait for the wind.

Tiny strands of silk trail from the
spider's silk glands, and when the wind
catches them, it whisks the spiders into
the air. This is called ballooning. The
lucky spiderlings land on plants where
they start making their own webs.

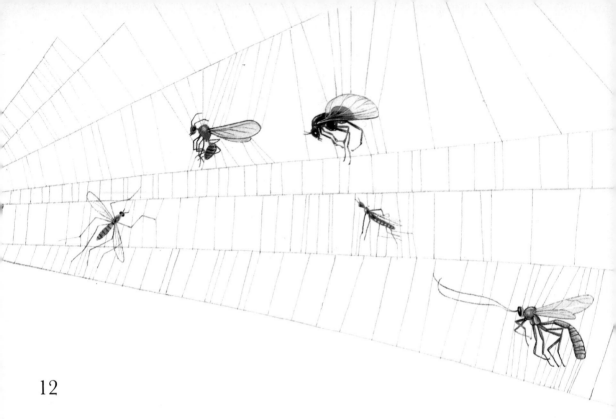

The baby spider never has a lesson on
web-making, but it knows just what to do.
It first makes a silken bridge between two
supports, often by letting a strand of silk
drift in the air until it catches on a nearby
branch. It then hangs its web from this
bridge.

You can see the bridge line at the top of
this photograph. The first web is small,
but it is big enough to catch small flies
and the spider grows fat on them.

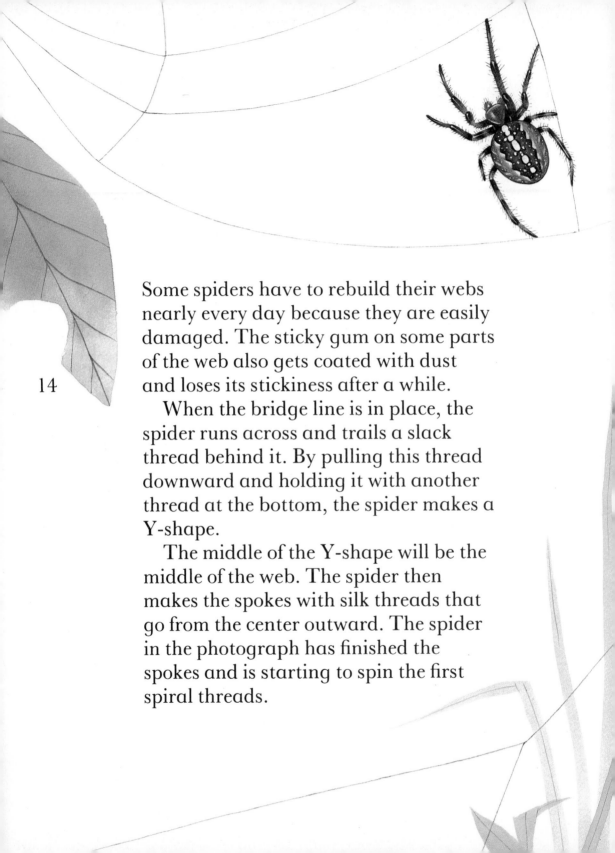

Some spiders have to rebuild their webs nearly every day because they are easily damaged. The sticky gum on some parts of the web also gets coated with dust and loses its stickiness after a while.

When the bridge line is in place, the spider runs across and trails a slack thread behind it. By pulling this thread downward and holding it with another thread at the bottom, the spider makes a Y-shape.

The middle of the Y-shape will be the middle of the web. The spider then makes the spokes with silk threads that go from the center outward. The spider in the photograph has finished the spokes and is starting to spin the first spiral threads.

14

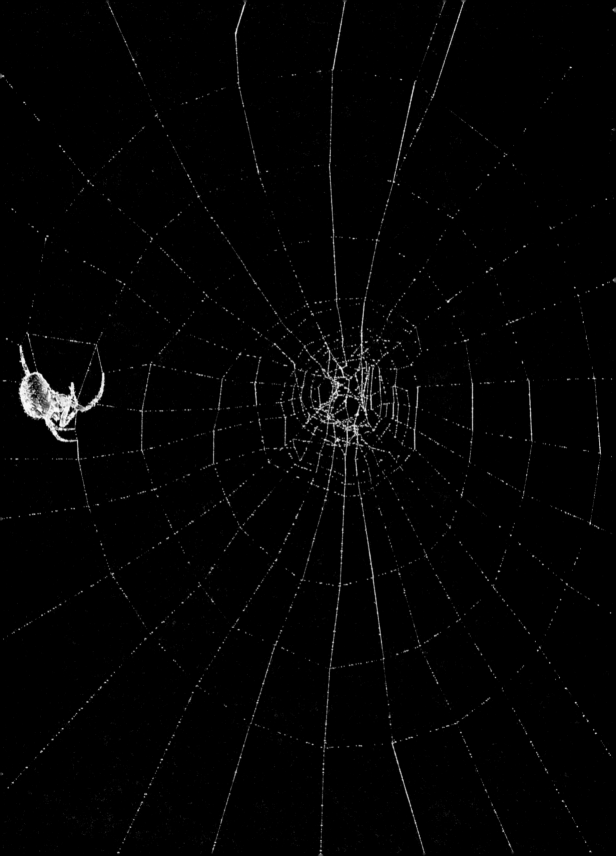

16

The first spiral threads are not sticky: they are like scaffolding for the spider to walk on while it finishes the web. The spider in the photograph is laying down the final spiral, made up of sticky silk that will trap the spider's prey. You can see the silk thread coming out from the tip of the spider's body.

The spider uses its legs to position the silk in just the right places. As the silk stretches, its coating of gum breaks up into tiny droplets. When dew clings to these droplets, the web shines like silver.

This fully-grown female spider has finished her web and waits for food to arrive. If you look carefully you will see that she is sitting on her web.

But the spider doesn't always wait in the middle of her web. Often she hides under a nearby leaf. She always holds a strand of silk connected to the web so that no matter where she is she knows when an insect enters the web.

Notice the white cross on the spider's back. The garden spider is sometimes called the cross spider.

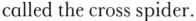

A wasp has flown into the spider's web and is struggling to get out. It has already broken part of the web, but the spider's sticky silk is very strong. Soon the spider will dash across the web to see what she has caught.

The spider's hooked claws help her grip the web and an oily liquid on her feet prevents her from sticking to the silk. She isn't frightened of the wasp, although it is almost as big as she is. She is too fast for the wasp to sting her.

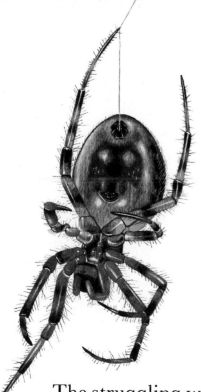

The struggling wasp has severely
damaged the web, but there is certainly
no escape for the wasp now. The spider
paralyzes the wasp with her poisonous
fangs. Then she wraps it in a wide band
of silk to tie it up. She spins the wasp
around with her front legs while pulling
the silk from her body with her back
legs.

This silk is different from the silks
used for the web and the egg cocoon.

The spider fixes the wasp securely to what remains of her web and settles down to eat her meal. Her mouth is small and she cannot take in any solid food. She pumps saliva all over the food and crushes it with the base of her fangs. She then waits until the saliva has dissolved the food and sucks it up into her stomach.

After her meal she has to make a new web. A large web may contain over 60 feet of silk connected by hundreds of joints, but the spider can finish the job in less than an hour.

The male garden spider is much smaller than the female and he has to be very careful when he goes courting. He signals to the female by pulling on her web, but it's dangerous for him as she might eat him.

After mating, the female spider will soon be ready to lay the eggs that are swelling up inside her and making her very fat.

It is now autumn. The spider lays her eggs and wraps them up in a silken bag. She is exhausted and very thin, and soon she will die.

In the spring the eggs will hatch and the spider's life story will begin again. Some of the baby spiders may grow up in the first summer, especially if the weather is warm. Garden spiders usually take about 18 months to reach their full size.

In the winter many spiders are dormant, or asleep.

Fascinating facts

The world's largest spiders are the Goliath bird-eating spiders of tropical South America. Some of these hairy spiders have bodies 3½ inches long and their legs can stretch across a dinner plate. They catch birds and mice as well as large insects.

The spitting spider is a small spider that catches its prey by spitting glue at it to stick it down.

The water spider makes a thimble-shaped web under the water – but not for catching prey. It fills the

Goliath bird-eating spider

web with air bubbles carried down from the surface and then sits in it. The spider then darts out to catch any small animal that comes along.

Gladiator spiders make small webs and throw them over passing insects.

Garden spiders and many other web-spinning spiders eat their old webs before making new ones, so the old silk is not wasted.

Undisturbed grassland may have as many as 500 spiders to every ten square feet.

Crab spiders sit in flowers and wait for insects to arrive. Some of them change their colors to match different flowers, so the insects don't notice them until it's too late.

Crab spider

Index